I0476198

Variants of Interaction Flow Diagrams

-- The Structure-Behavior Coalescence Approach --

William S. Chao

2

Structure-Behavior Coalescence

| Systems Architecture | = | Systems Structure | + | Systems Behavior |

CONTENTS

ABOUT THE AUTHOR

Dr. William S. Chao is the CEO & founder of SBC Architecture International®. SBC (Structure-Behavior Coalescence) architecture is a systems architecture which demands the integration of systems structure and systems behavior of a system. SBC architecture applies to hardware architecture, software architecture, enterprise architecture, knowledge architecture, and thinking architecture. The core theme of SBC architecture is: Architecture = Structure + Behavior.

William S. Chao received his bachelor degree (1976) in telecommunication engineering and master degree (1981) in information engineering, both from the National Chiao-Tung University, Taiwan. From 1976 till 1983, he worked as an engineer at Chung-Hwa Telecommunication Company, Taiwan.

William S. Chao received his master degree (1985) in information science and Ph.D. degree (1988) in information science, both from the University of Alabama at Birmingham, USA. From 1988 till 1991, he worked as a computer scientist at GE Research and Development Center, Schenectady, New York, USA.

Dr. William S. Chao has been teaching at National Sun Yat-

Sen University, Taiwan since 1992 and now serves as the president of Association of Enterprise Architects, Taiwan Chapter. His research covers: systems architecture, hardware architecture, software architecture, enterprise architecture, knowledge architecture, and thinking architecture.

PART I: CHANNEL-BASED INTERACTION FLOW DIAGRAMS

Channels

Channels are a model for agent communication. An agent may provide many channels, as shown in the figure below.

A channel may contain several input parameters (e.g. i_1, i_2) and output parameters (e.g. o_1, o_2).

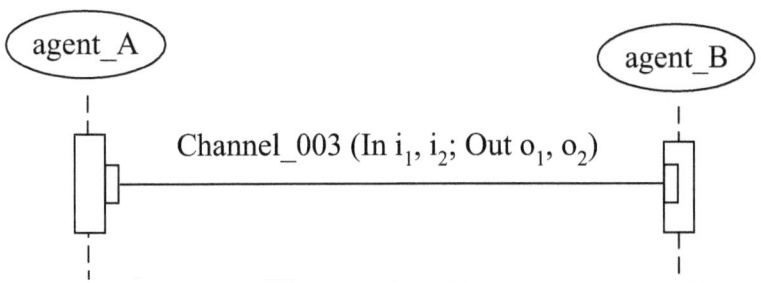

A channel formula is used to completely describe a channel. A channel formula includes a) channel name, b) input parameters (e.g. $i_1, i_2, \ldots, i_m$), and c) output parameters (e.g. $o_1, o_2, \ldots, o_n$).

Channel_Name (In $i_1, i_2, \ldots, i_m$; Out $o_1, o_2, \ldots, o_n$)

Channel-Based Value-Passing Interactions

An interaction represents an indivisible and instantaneous handshake or communication between two agents. In the channel-based approach, the caller agent (either external environment's actor or component) interacts with the callee agent (component) through the channel interaction.

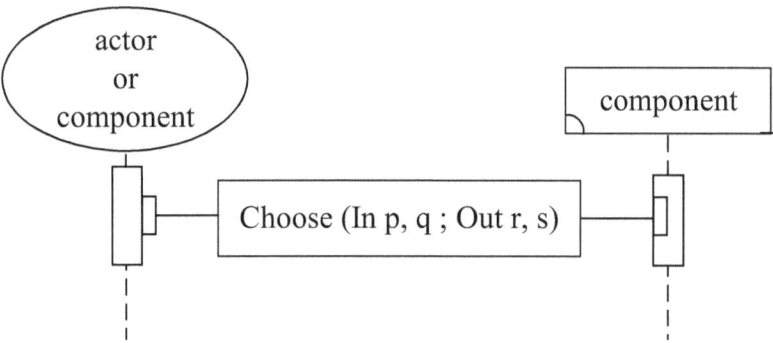

The caller agent owns the "calling port" of the interaction. In this case, the calling port is " $\overline{\text{Choose (In p, q; Out r, s)}}$ " and its conduct is to assist the caller agent to output a value to each of the "p" and "q" variables (of the "Choose" channel), and input a value from each of the "r" and "s" variables (of the "Choose" channel).

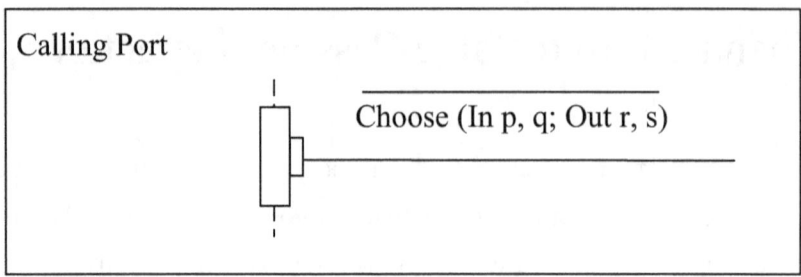

The caller agent together with the "calling port" is named the "calling action.

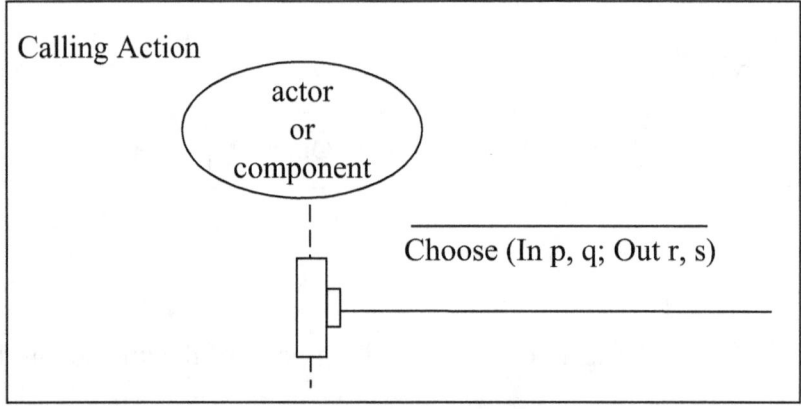

The callee agent owns the "called port" of the interaction. In this case, the called port is "Choose (In p, q; Out r, s)" and its conduct is to assist the callee agent to input a value from each of the "p" and "q" variables (of the "Choose" channel), and output a value to each of the "r" and "s" variables (of the "Choose" channel).

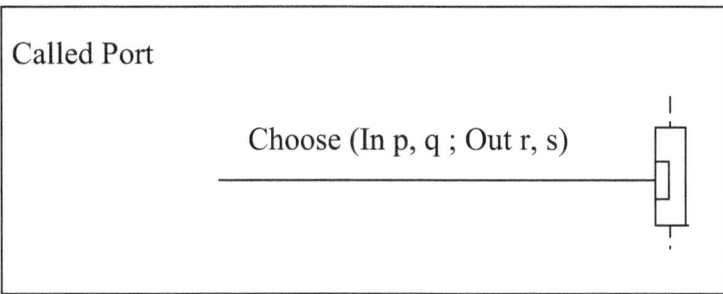

The callee agent together with the "called port" is named the "called action".

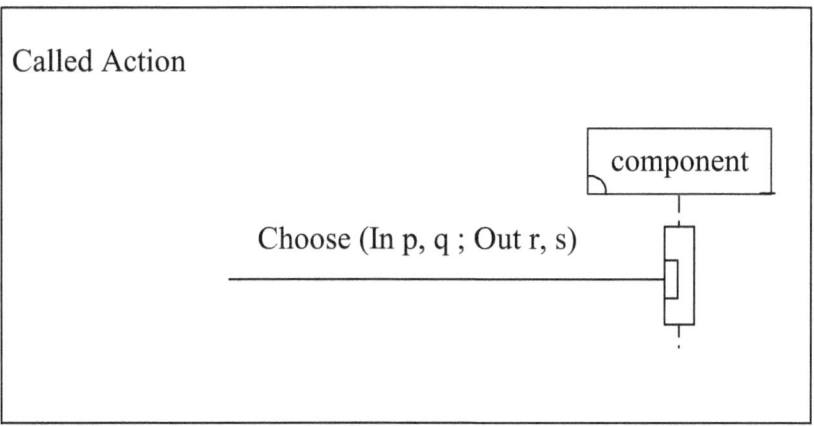

In order to simplify the channel-based interaction diagram, we will redraw it as follows.

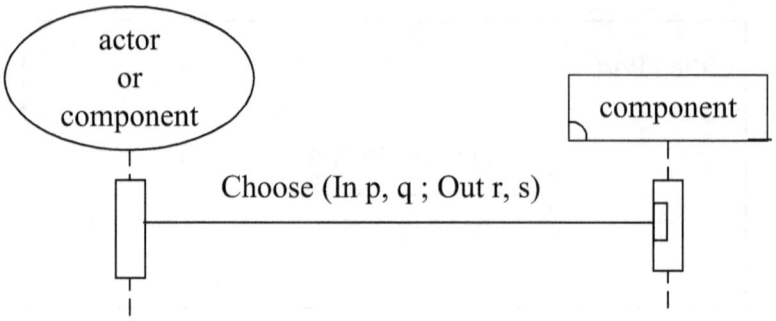

Or we can draw the channel-based interaction diagram as follows.

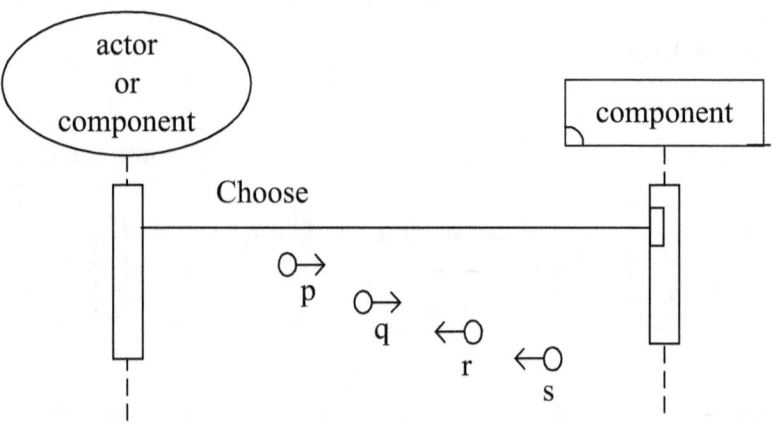

Formal Description of a Channel-Based Interaction

We formally describe a channel-based interaction as a 4-tuple INTERACTION = <caller_actor_or_component, channel_name, i/o_parameters, callee_component>, where "caller_actor_or_component" stands for the name of a caller actor or component, "channel_name" stands for the name of a channel, "i/o_parameters" stands for a 2-tuple of <input_parameters, output_parameters> where "input_parameters" stands for a set of input parameters and "output_parameters" stands for a set of output parameters, and "callee_component" stands for the name of a callee component.

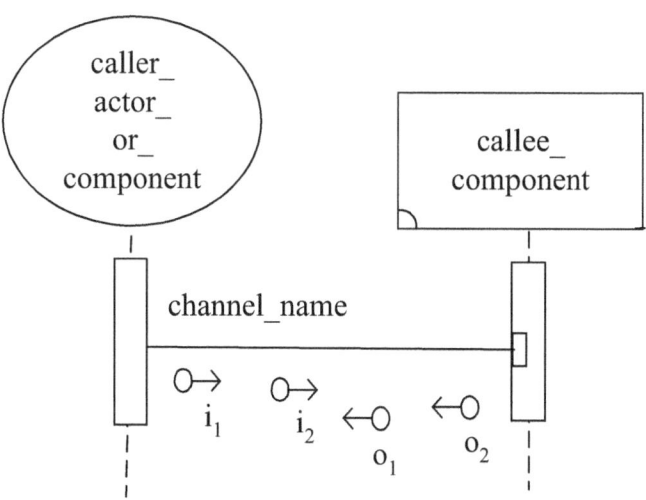

PART II: OPERATION-BASED INTERACTION FLOW DIAGRAMS

Operations

An operation provided by each component represents a procedure, or method, or function of the component.

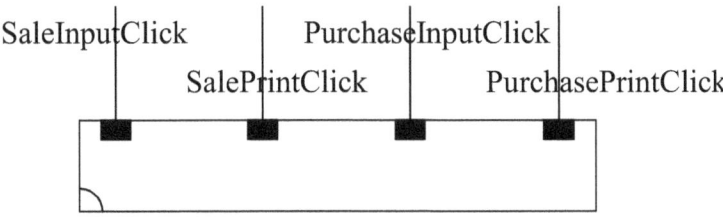

An operation may contain several input parameters (e.g. i_1, i_2) and output parameters (e.g. o_1, o_2).

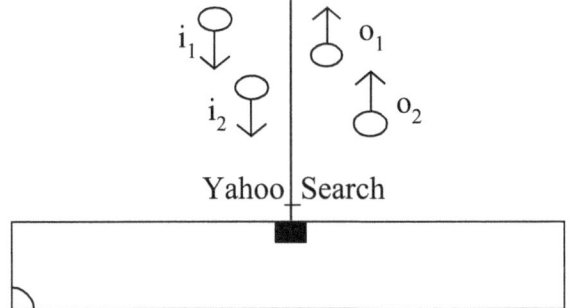

An operation formula is used to completely describe an operation. An operation formula includes a) operation name, b) input parameters (e.g. i_1, i_2, ..., i_m), and c) output parameters (e.g. o_1, o_2, ..., o_n).

Operation_Name (In i_1, i_2, ..., i_m ; Out o_1 , o_2, ..., o_n)

Operation-Based Value-Passing Interactions

An interaction represents an indivisible and instantaneous handshake or communication between two agents. In the operation-based approach, the caller agent (either external environment's actor or component) interacts with the callee agent (component) through the operation call or operation return interaction (also named as operation call or reply message).

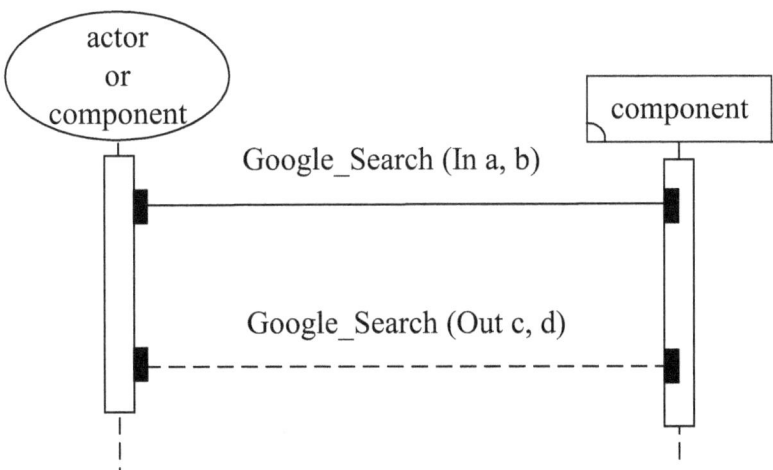

The caller agent owns the "calling port" of the interaction. In the operation call interaction (also known as operation call message) case, the calling port is " $\overline{\text{Google\_Search (In a, b)}}$,, and its conduct is to assist the caller agent to output a value to each of the "a" and "b" variables (of the "Google_Search" operation).

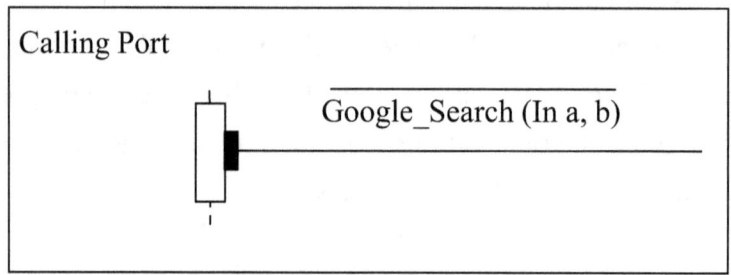

The caller agent together with the "calling port" is named the "calling action".

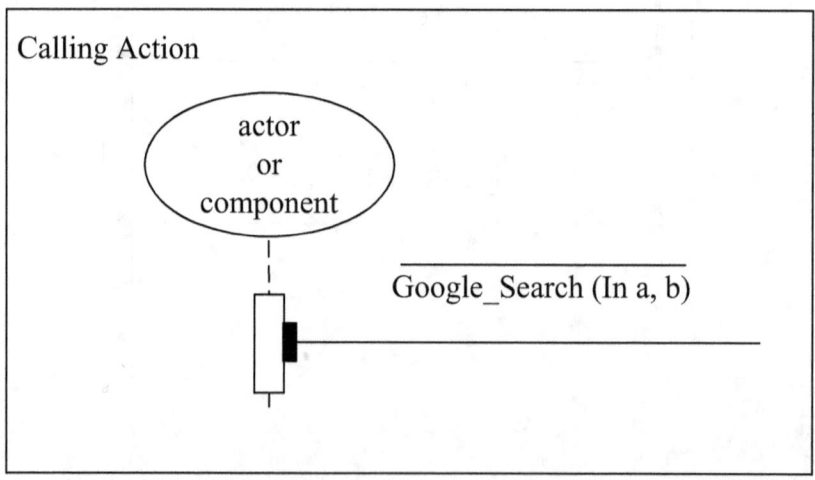

In the operation return interaction (also known as operation reply message) case, the calling port is " <u>Google_Search (Out c, d)</u> " and its conduct is to assist the caller agent to input a value from each of the "c" and "d" variables (of the "Google_Search" operation).

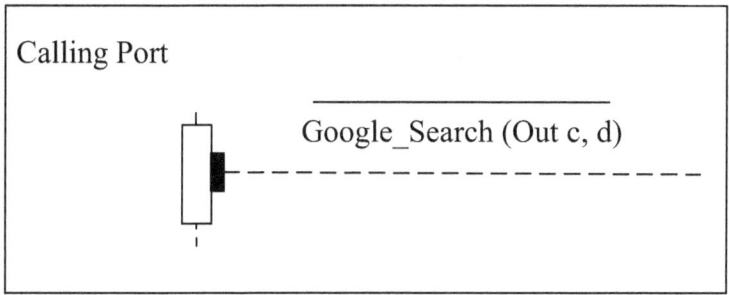

The caller agent together with the "calling port" is named the "calling action"

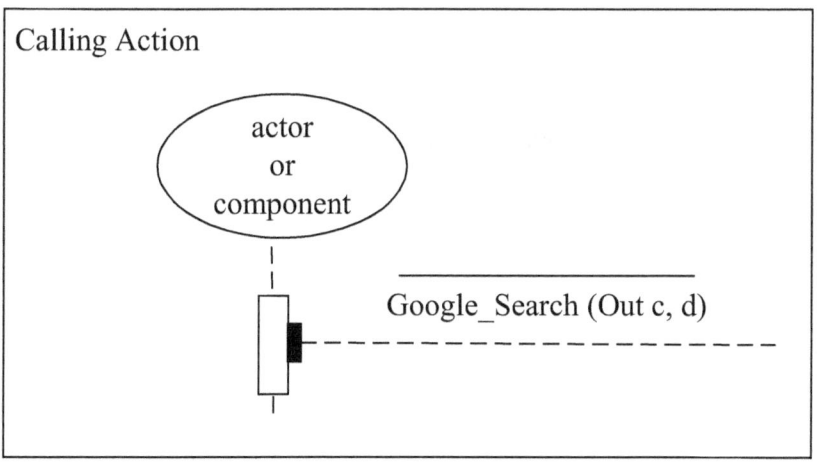

The callee agent owns the "called port" of the interaction. In the operation call interaction case, the called port is "Google_Search (In a, b)" and its conduct is to assist the callee agent to input a value from each of the "a" and "b" variables (of the "Google_Search" operation).

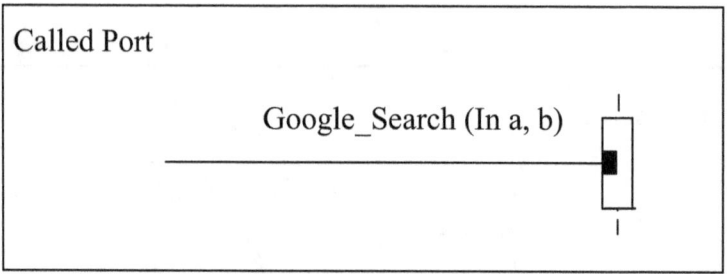

The callee agent together with the "called port" is named the "called action.

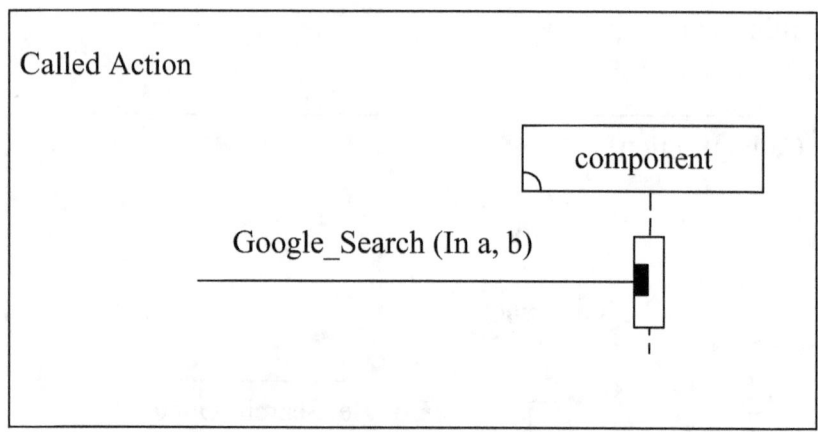

In the operation return interaction case, the called port is "Google_Search (Out c, d)" and its conduct is to assist the callee agent to output a value to each of the "c" and "d" variables (of the "Google_Search" operation).

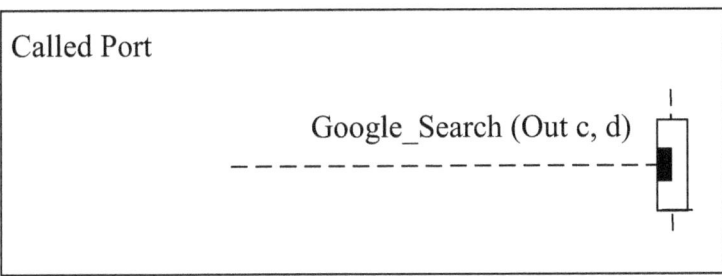

The callee agent together with the "called port" is named the "called action".

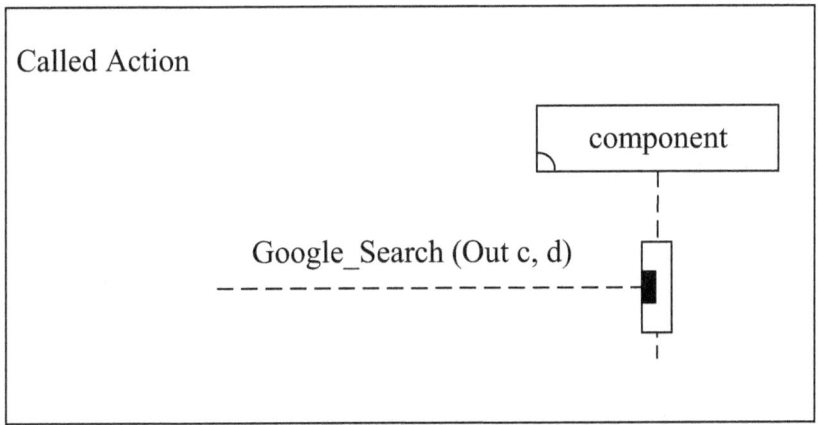

In order to simplify the operation-based interaction diagram, we will redraw it as follows.

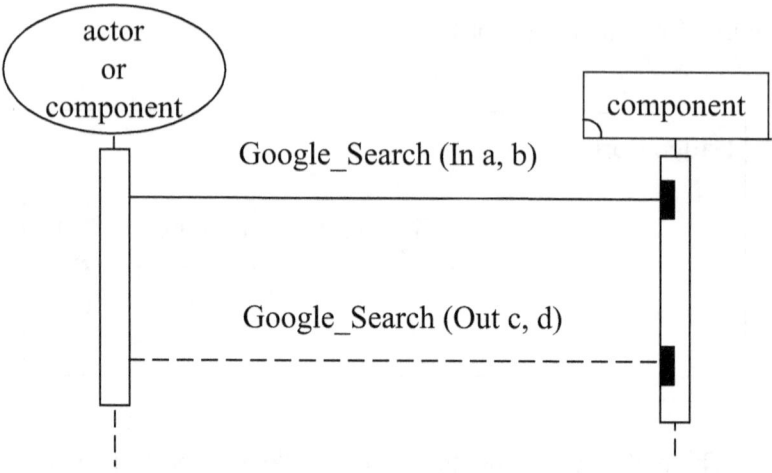

Or we can draw the operation-based interaction diagram as follows.

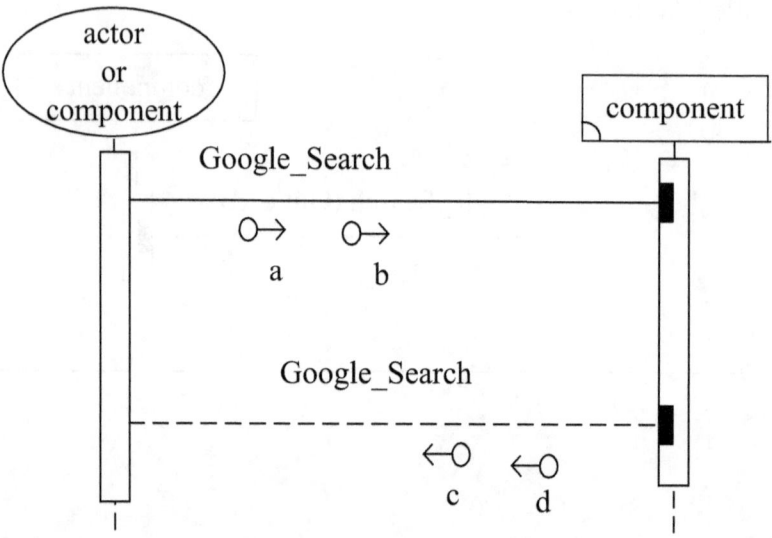

Formal Description of an Operation-Based Interaction

We formally describe an operation-based interaction as a 5-tuple INTERACTION = <operation_call_or_return, caller_actor_or_component, operation_name, i/o_parameters, callee_component>, where "operation_call_or_return" stands for an OPERATION_CALL or OPERATION_RETURN tag, "caller_actor_or_component" stands for the name of a caller actor or component, "operation_name" stands for the name of an operation, "i/o_parameters" stands for a 2-tuple of <input_parameters, output_parameters> where "input_parameters" stands for a set of input parameters and "output_parameters" stands for a set of output parameters, and "callee_component" stands for the name of a callee component.

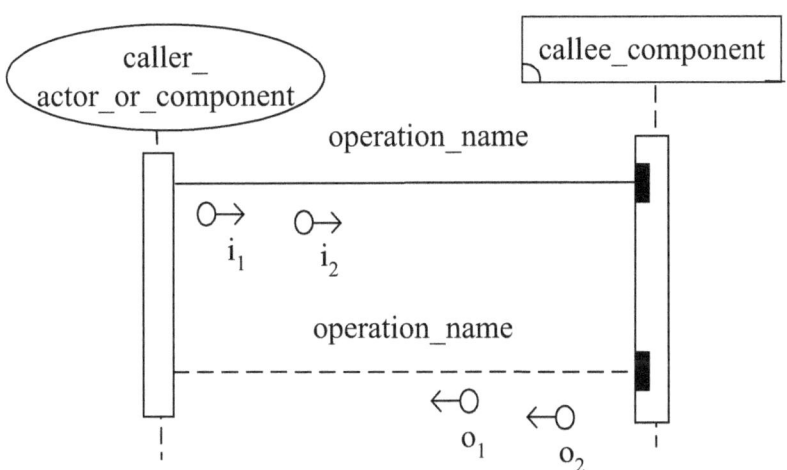

PART III: INTERACTION FLOW DIAGRAMS DEFINE THE SYSTEMS ARCHITECTURE

Interactions among Components and Actors to Draw Forth the Systems Behavior

In a system, if the components, and among them and the external environment's actors to interact (or handshake), these interactions will draw forth the systems behavior.

We conclude that "interaction" plays an important factor in integrating the systems structure and systems behavior for a system.

The overall behavior of a system consists of many individual behaviors. Each individual behavior represents an execution path. We use an interaction flow diagram (IFD) to demonstrate this individual behavior.

Formal Description of an Interaction Flow Diagram

An interaction flow diagram (IFD) is constructed by a sequence of interactions among the components and external environment's actors.

We formally describe the xth interaction flow diagram (IFD) of the systems architecture as a sequence $IFD_x = (interaction_{xz})_{z=1 \text{ to } N}$, where "z" stands for the zth (z = 1 to N) interaction of this xth interaction flow diagram.

Or we can formally describe the xth interaction flow diagram as a sequence $IFD_x = (interaction_{x1}, interaction_{x2}, interaction_{x3},..., interaction_{xN})$.

Examples of Formal Description of an Interaction Flow Diagram

We suppose the example systems architecture consists of two interaction flow diagrams.

The first interaction flow diagram consists of two interactions.

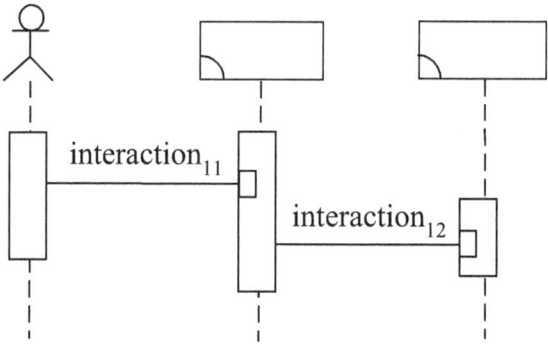

We formally describe this 1st interaction flow diagram as a sequence $IFD_1 = (interaction_{11}, interaction_{12})$.

The second interaction flow diagram consists of three interactions.

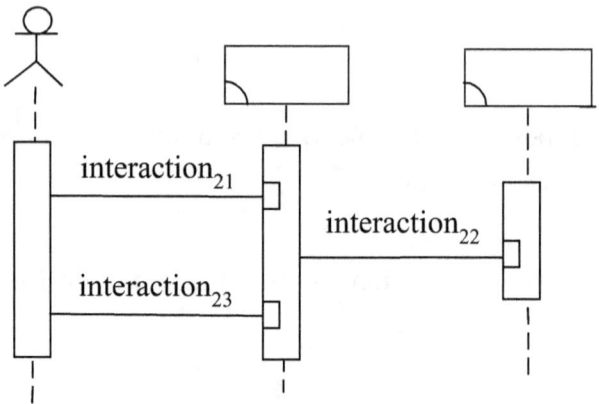

We formally describe this 2nd interaction flow diagram as a sequence $IFD_2 = (interaction_{21}, interaction_{22}, interaction_{23})$.

Collection of All Interaction Flow Diagrams Defines the Systems Architecture

The collection of all interaction flow diagrams defines the integration of systems structure and systems behavior (i.e. structure-behavior coalescence) of a system.

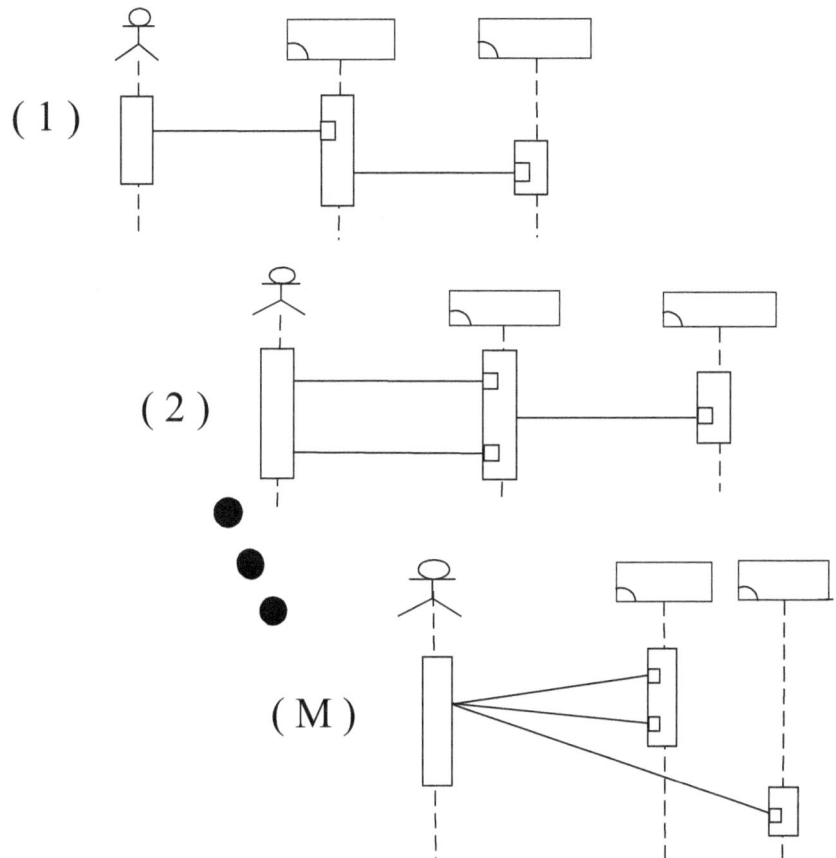

That is, the collection of all interaction flow diagrams defines the systems architecture.

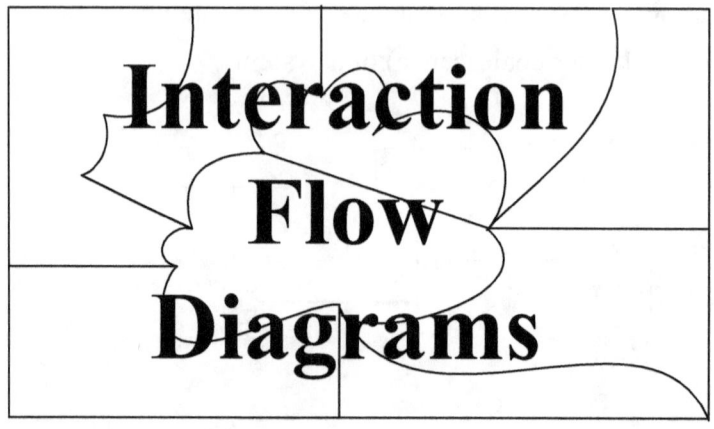

Systems architecture (SA) represents a knowledge repository of a system. Stakeholders can submit and acquire knowledge to and from this knowledge repository.

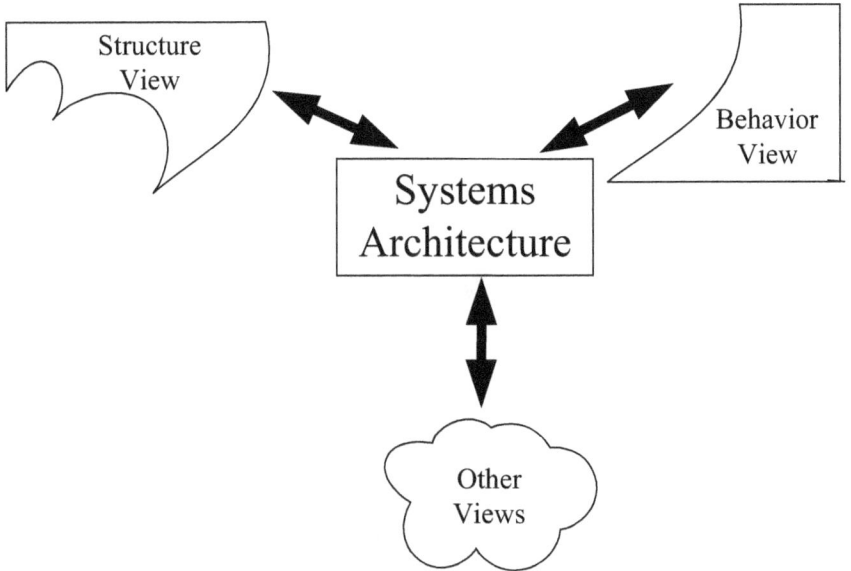

So, the collection of all interaction flow diagrams represents a knowledge repository of a system. Stakeholders can submit and acquire knowledge to and from this knowledge repository.

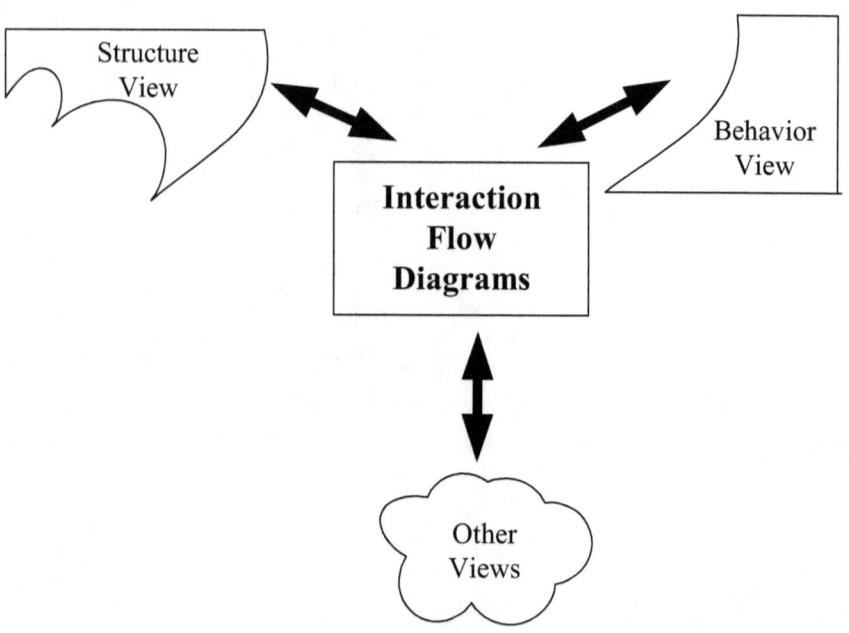

Formal Description of the Systems Architecture

Since the collection of all interaction flow diagrams defines the systems architecture, the systems architecture can be formally described as a M-tuple IFDTUPLE = $<IFD_x>_{x = 1 \text{ to } M}$, where "x" stands for the xth interaction flow diagram of this collection of all interaction flow diagrams.

Or we can formally describe the systems architecture as a M-tuple IFDTUPLE = $<IFD_1, IFD_2, IFD_3,..., IFD_M>$.

Examples of Formal Description of the Systems Architecture

We suppose the example systems architecture consists of two interaction flow diagrams.

The first interaction flow diagram consists of two interactions.

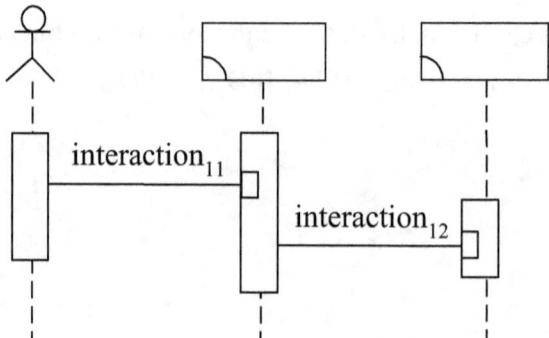

The second interaction flow diagram consists of three interactions.

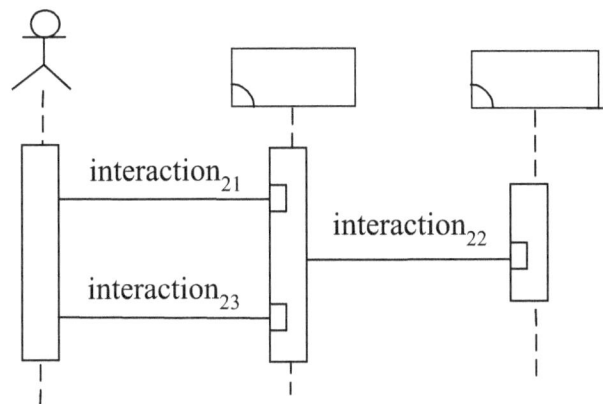

We formally describe this systems architecture as a 2-tuple IFDTUPLE = <IFD$_1$, IFD$_2$>.

Or we formally describe it as a 2-tuple IFDTUPLE = <(interaction$_{11}$, interaction$_{12}$), (interaction$_{21}$, interaction$_{22}$, interaction$_{23}$)>.

PART IV: FIRST EXAMPLE -- MULTI-TIER PERSONAL DATA SYSTEM

Channel-Based Systems Architecture of the Multi-Tier Personal Data System

The collection of all channel-based interaction flow diagrams defines the systems architecture. The overall behavior of the *Multi-Tier Personal Data System* includes two behaviors: *AgeCalculation* and *OverweightCalculation*. Each of them is described by an individual channel-based IFD.

The channel-based IFD of the *AgeCalculation* behavior is shown below. First, actor *Pupil* interacts with the *MTPDS_GUI* component through the *Calculate_AgeClick_Call* channel interaction, carrying the *Social_Security_Number* input parameter. Next, component *MTPDS_GUI* interacts with the *Age_Logic* component through the *Calculate_Age_Call* channel interaction, carrying the *Social_Security_Number* input parameter. Continuingly, component *Age_Logic* interacts with the *Personal_Database* component through the *Sql_DateOfBirth_Select* channel interaction, carrying the *Social_Security_Number* input parameter and the *query_DateOfBirth* output parameter. Repeatedly, component *MTPDS_GUI* interacts with the *Age_Logic* component through the *Calculate_Age_Return* channel interaction, carrying the *Age* output parameter. Finally, actor *Pupil* interacts with the *MTPDS_GUI* component through the *Calculate_AgeClick_Return* channel interaction, carrying the *Age* output parameter.

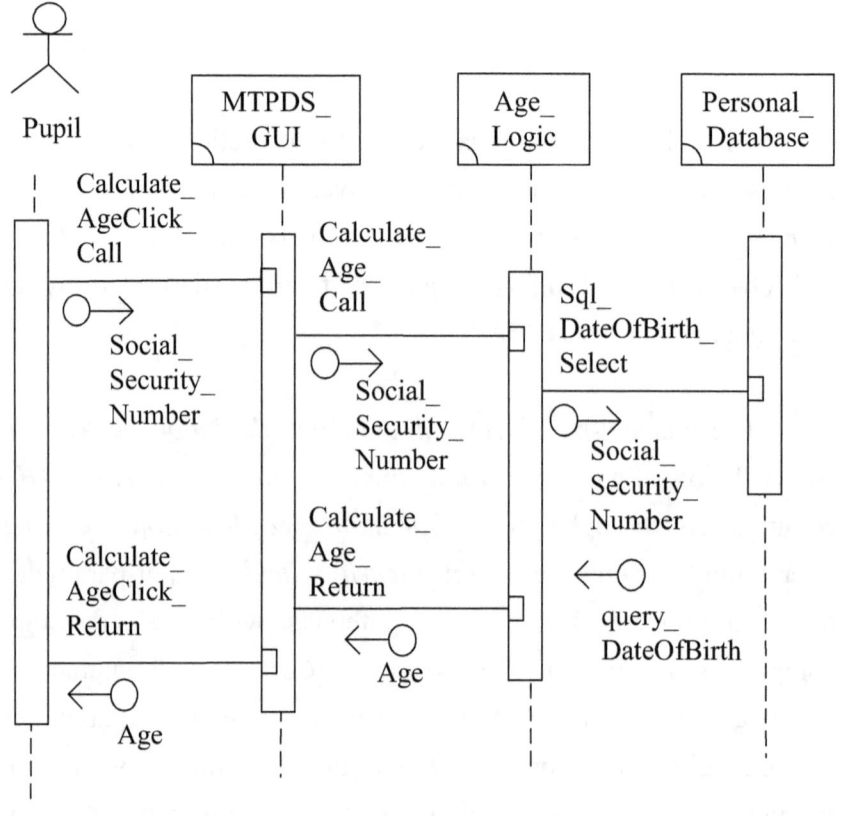

The channel-based IFD of the *OverweightCalculation* behavior is shown below. First, actor *Pupil* interacts with the *MTPDS_GUI* component through the *Calculate_OverweightClick_Call* channel interaction, carrying the *Social_Security_Number* input parameter. Next, component *MTPDS_GUI* interacts with the *Overweight_Logic* component through the *Calculate_Overweight_Call* channel interaction, carrying the *Social_Security_Number* input parameter. Continuingly, component *Overweight_Logic* interacts with the *Personal_Database* component through the *Sql_SexHeightWeight_Select* channel interaction, carrying the *Social_Security_Number* input parameter and the *query_SexHeightWeight* output parameter. Repeatedly, component *MTPDS_GUI* interacts with the *Overweight_Logic* component through the *Calculate_Overweight_Return* channel interaction, carrying the *Overweight* output parameter. Finally, actor *Pupil* interacts with the *MTPDS_GUI* component through the *Calculate_OverweightClick_Return* channel interaction, carrying the *Overweight* output parameter.

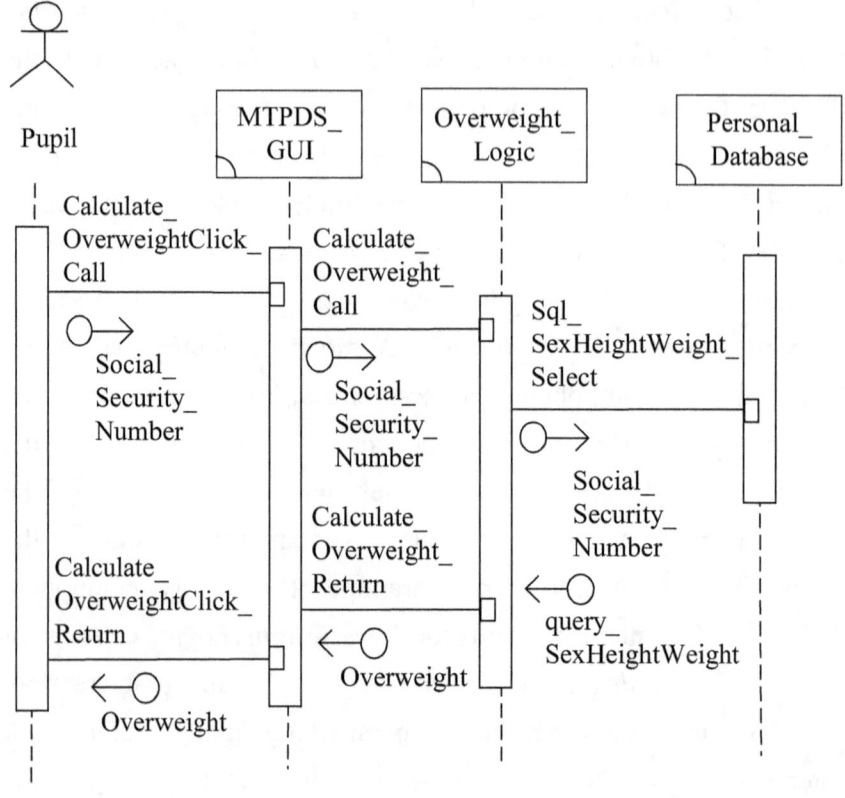

Operation-Based Systems Architecture of the Multi-Tier Personal Data System

The collection of all operation-based interaction flow diagrams defines the systems architecture. The overall behavior of the *Multi-Tier Personal Data System* includes two behaviors: *AgeCalculation* and *OverweightCalculation*. Each of them is described by an individual operation-based IFD.

The operation-based IFD of the *AgeCalculation* behavior is shown below. First, actor *Pupil* interacts with the *MTPDS_GUI* component through the *Calculate_AgeClick* operation call interaction, carrying the *Social_Security_Number* input parameter. Next, component *MTPDS_GUI* interacts with the *Age_Logic* component through the *Calculate_Age* operation call interaction, carrying the *Social_Security_Number* input parameter. Continuingly, component *Age_Logic* interacts with the *Personal_Database* component through the *Sql_DateOfBirth_Select* operation call interaction, carrying the *Social_Security_Number* input parameter and the *query_DateOfBirth* output parameter. Repeatedly, component *MTPDS_GUI* interacts with the *Age_Logic* component through the *Calculate_Age* operation return interaction, carrying the *Age* output parameter. Finally, actor *Pupil* interacts with the *MTPDS_GUI* component through the *Calculate_AgeClick* operation return interaction, carrying the *Age* output parameter.

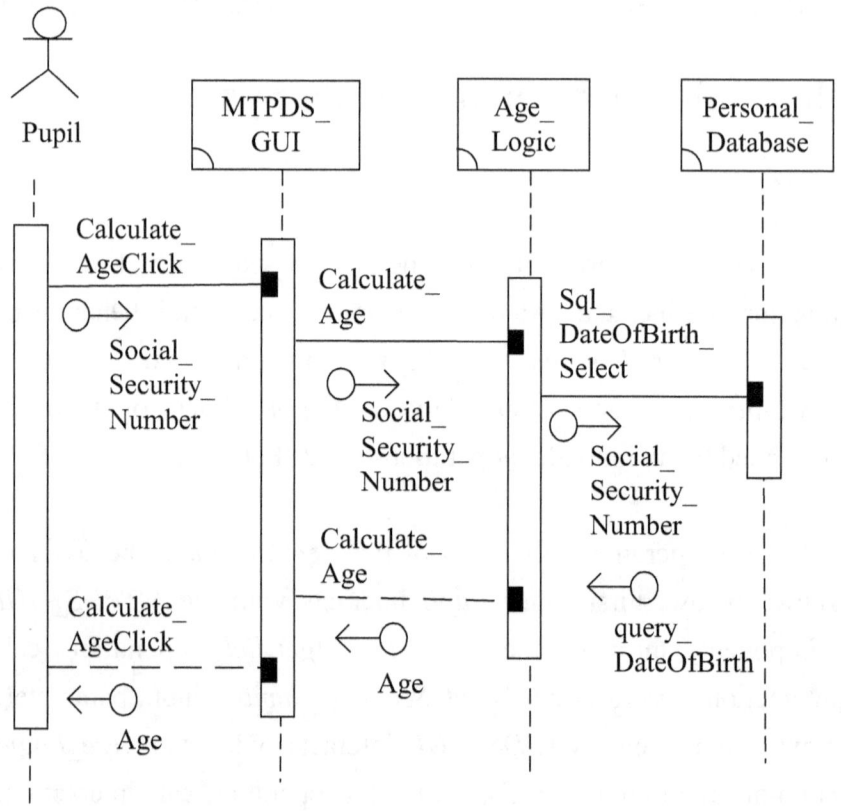

The operation-based IFD of the *OverweightCalculation* behavior is shown below. First, actor *Pupil* interacts with the *MTPDS_GUI* component through the *Calculate_OverweightClick* operation call interaction, carrying the *Social_Security_Number* input parameter. Next, component *MTPDS_GUI* interacts with the *Overweight_Logic* component through the *Calculate_Overweight* operation call interaction, carrying the *Social_Security_Number* input parameter. Continuingly, component *Overweight_Logic*

interacts with the *Personal_Database* component through the *Sql_SexHeightWeight_Select* operation call interaction, carrying the *Social_Security_Number* input parameter and the *query_SexHeightWeight* output parameter. Repeatedly, component *MTPDS_GUI* interacts with the *Overweight_Logic* component through the *Calculate_Overweight* operation return interaction, carrying the *Overweight* output parameter. Finally, actor *Pupil* interacts with the *MTPDS_GUI* component through the *Calculate_OverweightClick* operation return interaction, carrying the *Overweight* output parameter.

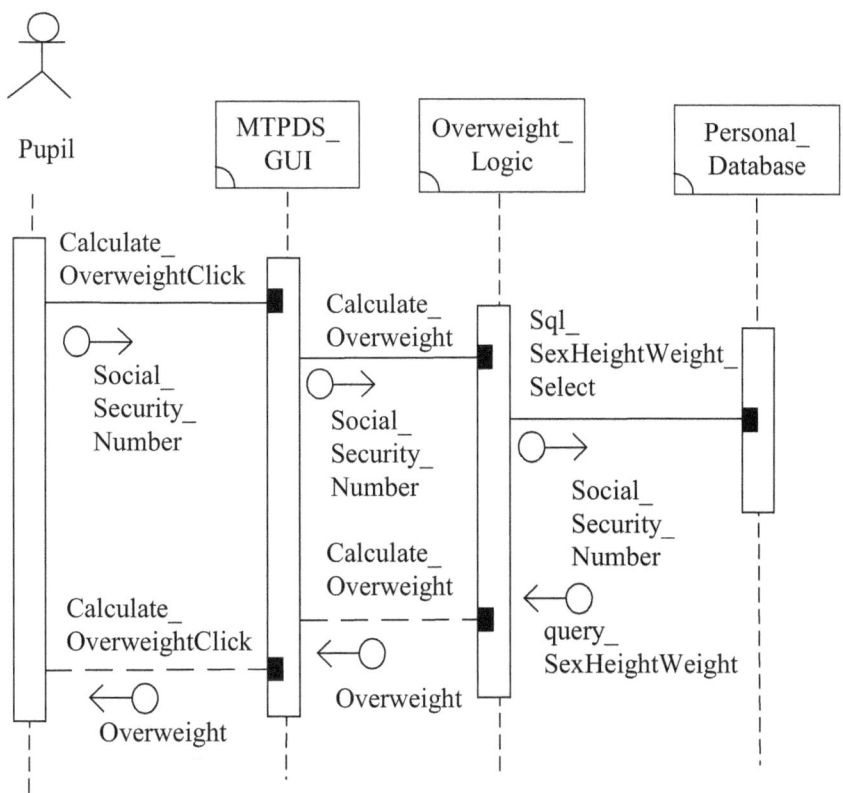

PART V: SECOND EXAMPLE --
SALE AND PURCHASE SYSTEM

Channel-Based Systems Architecture of the Sale and Purchase System

The collection of all channel-based interaction flow diagrams defines the systems architecture. The overall behavior of of the *Sale and Purchase System* includes four behaviors: *SaleInput*, *SalePrint*, *PurchaseInput*, and *PurchasePrint*. Each of them is described by an individual channel-based IFD.

The channel-based IFD of the *SaleInput* behavior is shown below. First, actor *Sales Clerk* interacts with the *SalePurchase_GUI* component through the *SaleInputClick* channel interaction. Next, component *SalePurchase_GUI* interacts with the *SaleInput_GUI* component through the *ShowModal* channel interaction. Continuingly, actor *Sales Clerk* interacts with the *SaleInput_GUI* component through the *SaleDataInput* channel interaction, carrying the *s_form* input parameter. Finally, component *SaleInput_GUI* interacts with the *SalePurchase_Database* component through the *Sql_s_insert* channel interaction, carrying the *s_query_1* input parameter.

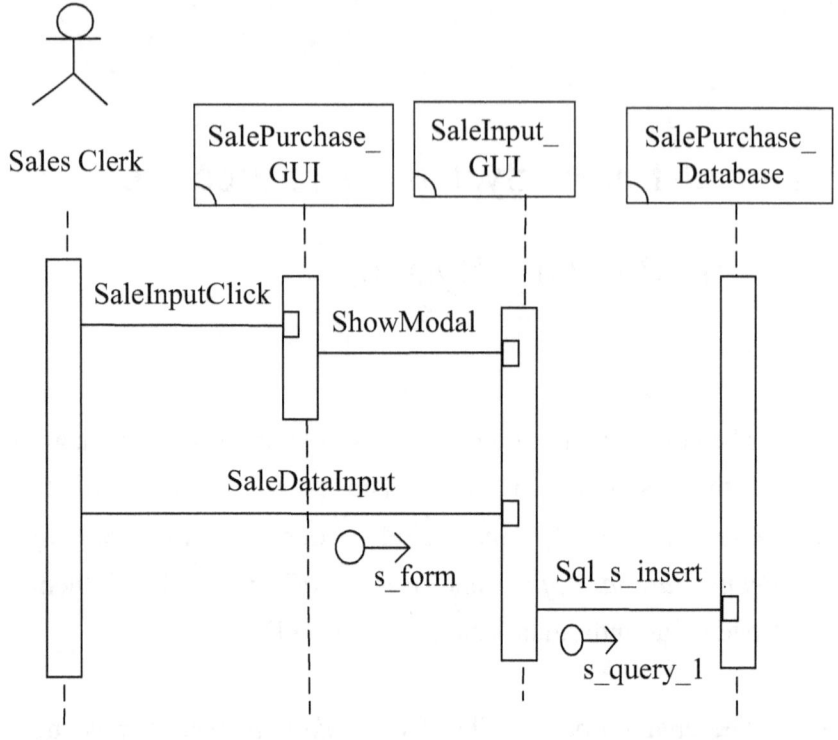

The channel-based IFD of the *SalePrint* behavior is shown below. First, actor *Sales Clerk* interacts with the *SalePurchase_GUI* component through the *SalePrintClick* channel interaction. Next, component *SalePurchase_GUI* interacts with the *SalePrint_GUI* component through the *ShowModal* channel interaction. Continuingly, actor *Sales Clerk* interacts with the *SalePrint_GUI* component through the *SalePrintButtonClick_Call* channel interaction, carrying the *sDate* and *sNo* input parameters. Continuingly, component *SalePrint_GUI* interacts with the *SalePurchase_Database* component through the *Sql_s_select* channel interaction, carrying the *sDate* and *sNo* input parameters and the *s_query_2* output parameter. Finally, actor *Sales Clerk*

interacts with the *SalePrint_GUI* component through the *SalePrintButtonClick_Return* channel interaction, carrying the *s_report* output parameter.

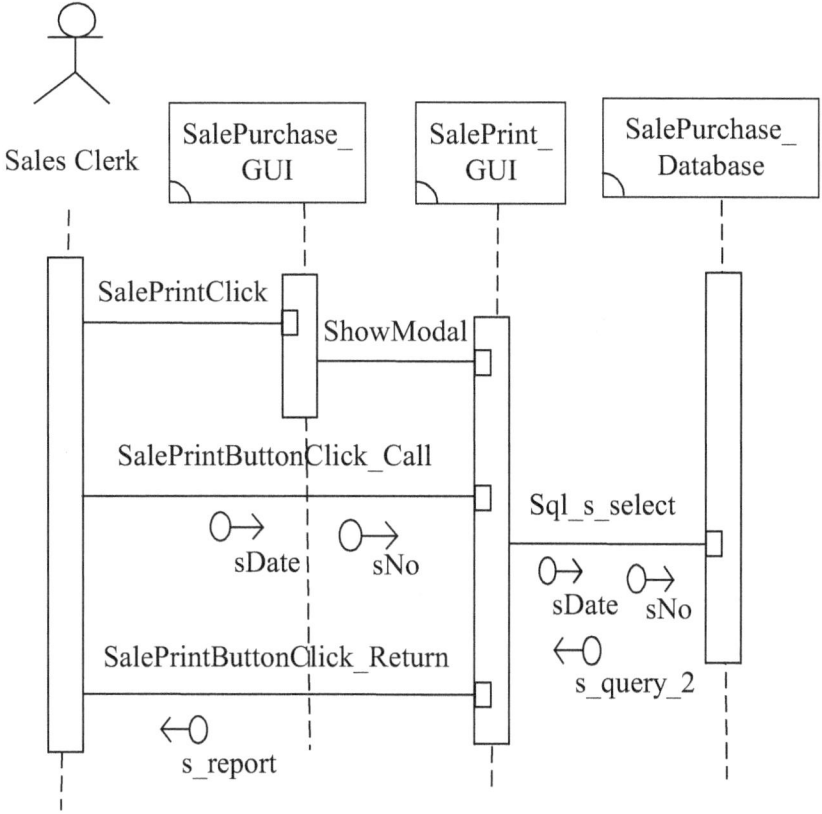

The channel-based IFD of the *PurchaseInput* behavior is shown below. First, actor *Purchase Clerk* interacts with the *SalePurchase_GUI* component through the *PurchaseInputClick* channel interaction. Next, component *SalePurchase_GUI* interacts with the *PurchaseInput_GUI* component through the *ShowModal* channel interaction. Continuingly, actor *Purchase Clerk* interacts with the *PurchaseInput_GUI* component through the

PurchaseDataInput channel interaction, carrying the *p_form* input parameter. Finally, component *PurchaseInput_GUI* interacts with the *SalePurchase_Database* component through the *Sql_p_insert* channel interaction, carrying the *p_query_1* input parameter.

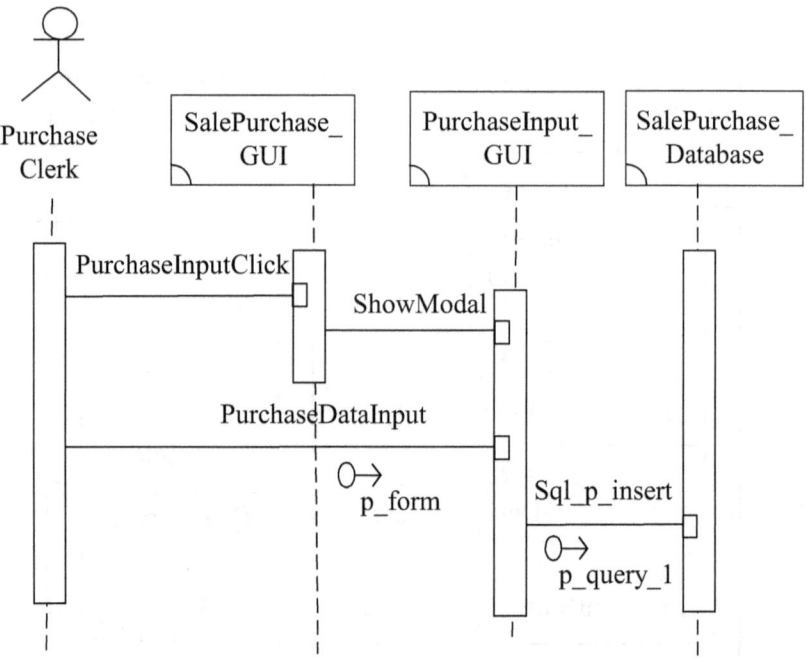

The channel-based IFD of the *PurchasePrint* behavior is shown below. First, actor *Purchase Clerk* interacts with the *SalePurchase_GUI* component through the *PurchasePrintClick* channel interaction. Next, component *SalePurchase_GUI* interacts with the *PurchasePrint_GUI* component through the *ShowModal* channel interaction. Continuingly, actor *Purchase Clerk* interacts with the *PurchasePrint_GUI* component through the *PurchasePrintButtonClick_Call* channel interaction, carrying the *pDate* and *pNo* input parameters. Continuingly, component *PurchasePrint_GUI* interacts with the *SalePurchase_Database*

component through the *Sql_p_select* channel interaction, carrying the *pDate* and *pNo* input parameters and the *p_query_2* output parameter. Finally, actor *Purchase Clerk* interacts with the *PurchasePrint_GUI* component through the *PurchasePrintButtonClick_Return* channel interaction, carrying the *p_report* output parameter.

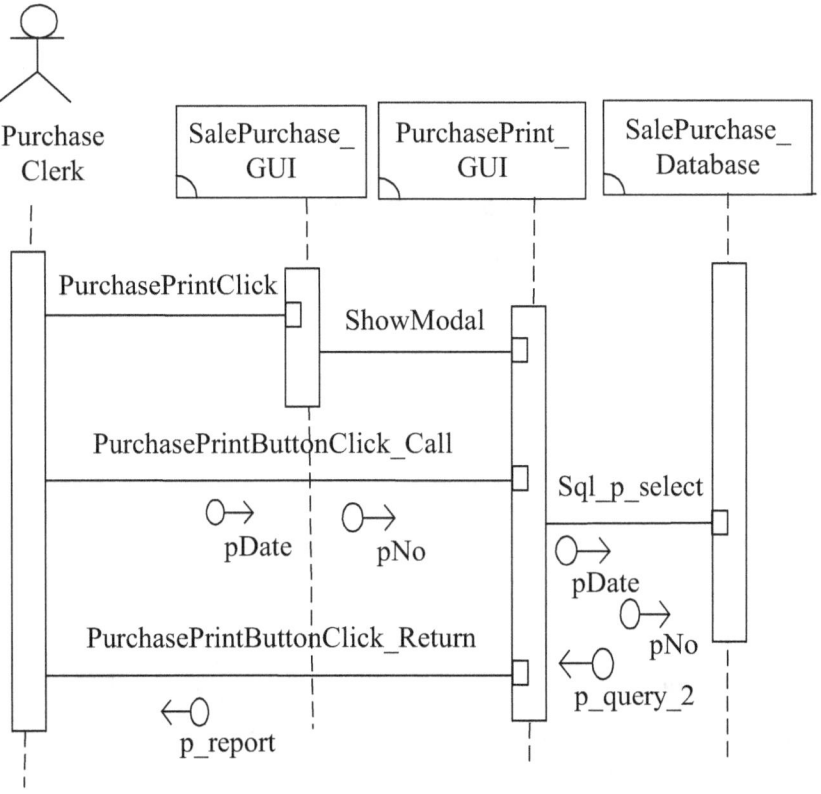

Operation-Based Systems Architecture of the Sale and Purchase System

The collection of all operation-based interaction flow diagrams defines the systems architecture. The overall behavior of of the *Sale and Purchase System* includes four behaviors: *SaleInput*, *SalePrint*, *PurchaseInput*, and *PurchasePrint*. Each of them is described by an individual operation-based IFD.

The operation-based IFD of the *SaleInput* behavior is shown below. First, actor *Sales Clerk* interacts with the *SalePurchase_GUI* component through the *SaleInputClick* operation call interaction. Next, component *SalePurchase_GUI* interacts with the *SaleInput_GUI* component through the *ShowModal* operation call interaction. Continuingly, actor *Sales Clerk* interacts with the *SaleInput_GUI* component through the *SaleDataInput* operation call interaction, carrying the *s_form* input parameter. Finally, component *SaleInput_GUI* interacts with the *SalePurchase_Database* component through the *Sql_s_insert* operation call interaction, carrying the *s_query_1* input parameter.

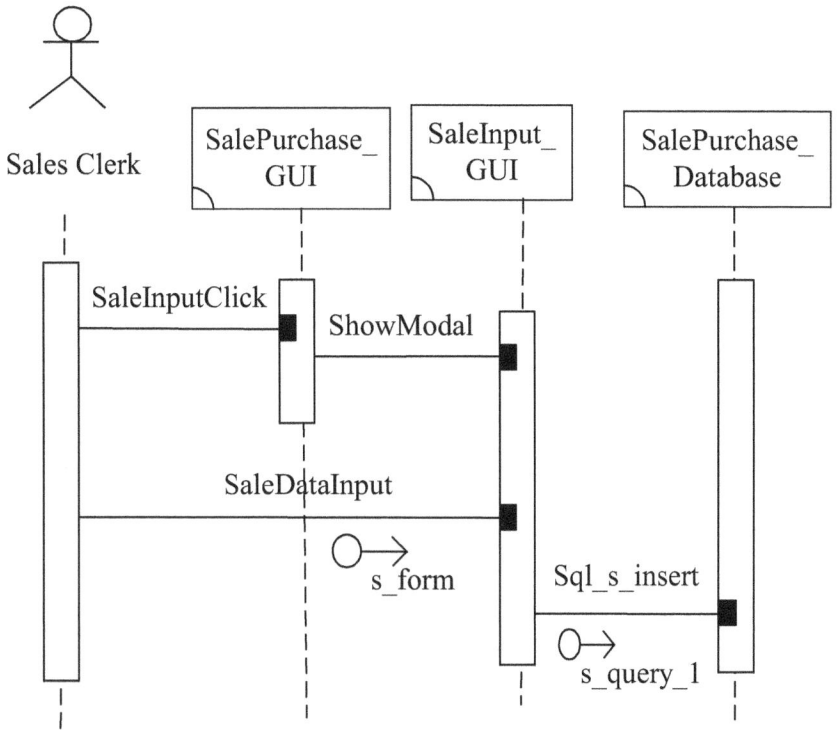

The operation-based IFD of the *SalePrint* behavior is shown below. First, actor *Sales Clerk* interacts with the *SalePurchase_GUI* component through the *SalePrintClick* operation call interaction. Next, component *SalePurchase_GUI* interacts with the *SalePrint_GUI* component through the *ShowModal* operation call interaction. Continuingly, actor *Sales Clerk* interacts with the *SalePrint_GUI* component through the *SalePrintButtonClick* operation call interaction, carrying the *sDate* and *sNo* input parameters. Continuingly, component *SalePrint_GUI* interacts with the *SalePurchase_Database* component through the *Sql_s_select* operation call interaction, carrying the *sDate* and *sNo* input parameters and the *s_query_2*

output parameter. Finally, actor *Sales Clerk* interacts with the *SalePrint_GUI* component through the *SalePrintButtonClick* operation return interaction, carrying the *s_report* output parameter.

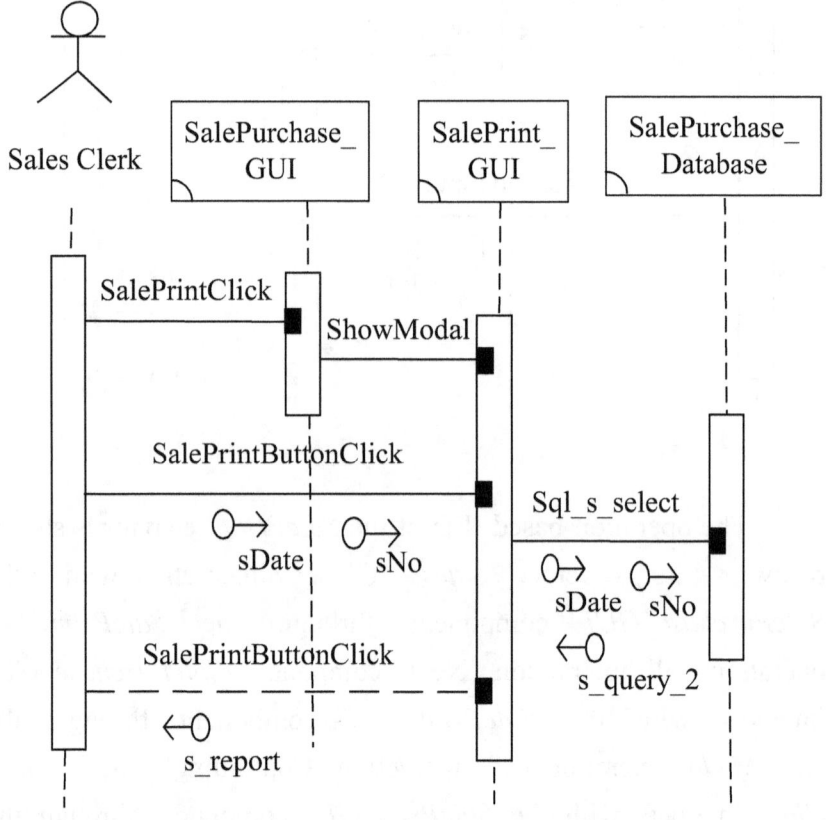

The operation-based IFD of the *PurchaseInput* behavior is shown below. First, actor *Purchase Clerk* interacts with the *SalePurchase_GUI* component through the *PurchaseInputClick* operation call interaction. Next, component *SalePurchase_GUI* interacts with the *PurchaseInput_GUI* component through the *ShowModal* operation call interaction. Continuingly, actor *Purchase Clerk* interacts with the *PurchaseInput_GUI* component through the *PurchaseDataInput* operation call interaction, carrying the *p_form* input parameter. Finally, component *PurchaseInput_GUI* interacts with the *SalePurchase_Database* component through the *Sql_p_insert* operation call interaction, carrying the *p_query_1* input parameter.

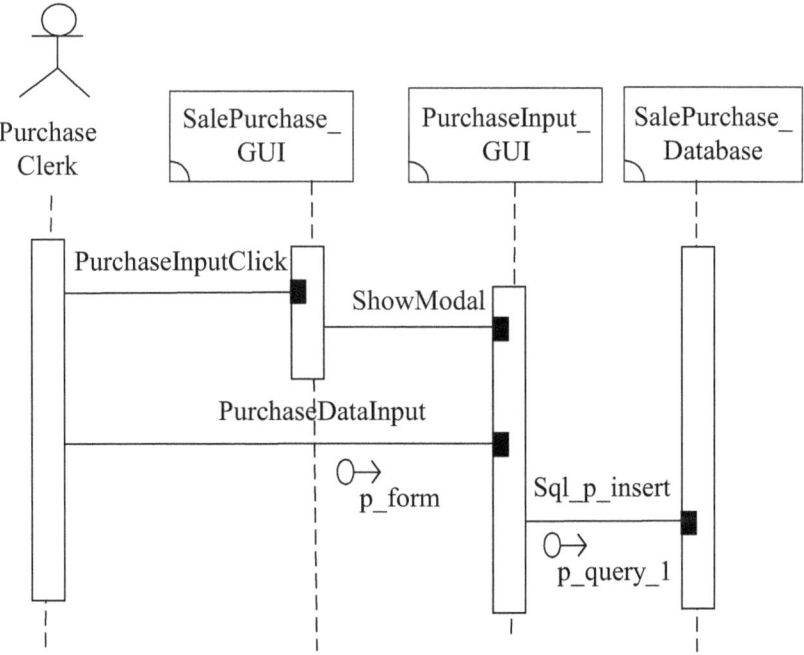

The operation-based IFD of the *PurchasePrint* behavior is

shown below. First, actor *Purchase Clerk* interacts with the *SalePurchase_GUI* component through the *PurchasePrintClick* operation call interaction. Next, component *SalePurchase_GUI* interacts with the *PurchasePrint_GUI* component through the *ShowModal* operation call interaction. Continuingly, actor *Purchase Clerk* interacts with the *PurchasePrint_GUI* component through the *PurchasePrintButtonClick* operation call interaction, carrying the *pDate* and *pNo* input parameters. Continuingly, component *PurchasePrint_GUI* interacts with the *SalePurchase_Database* component through the *Sql_p_select* operation call interaction, carrying the *pDate* and *pNo* input parameters and the *p_query_2* output parameter. Finally, actor *Purchase Clerk* interacts with the *PurchasePrint_GUI* component through the *PurchasePrintButtonClick* operation return interaction, carrying the *p_report* output parameter.

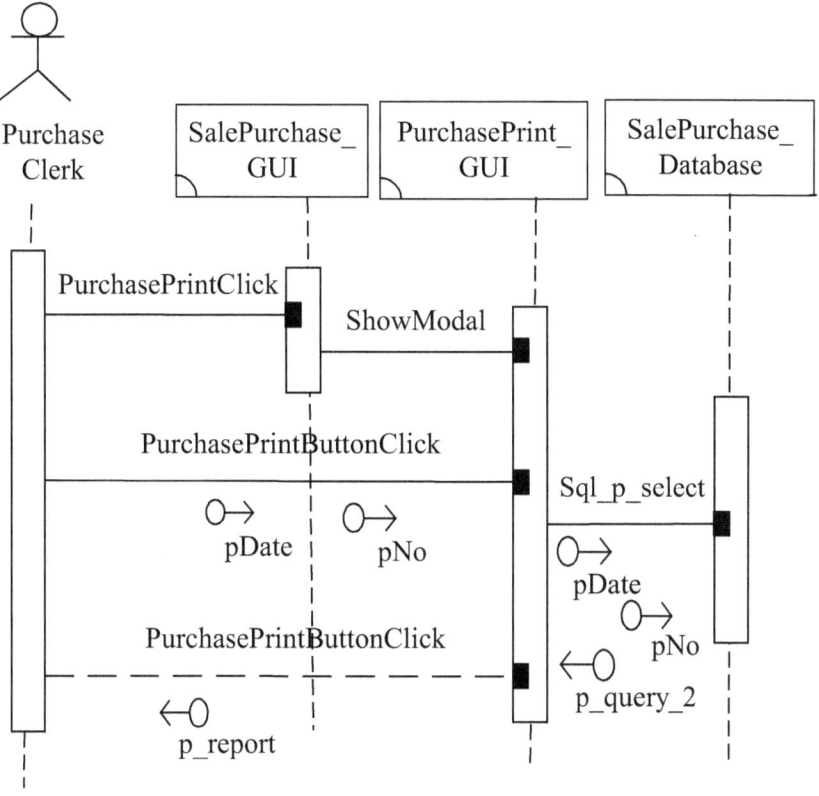

PART VI: THIRD EXAMPLE -- RESTAURANT

Channel-Based Systems Architecture of the Restaurant

The collection of all channel-based interaction flow diagrams defines the systems architecture. The overall behavior of the *Restaurant* includes two behaviors: *Ordering_a_Meal* and *Paying_the_Bill*. Each of them is described by an individual channel-based IFD.

The channel-based IFD of the *Ordering_a_Meal* behavior is shown below. First, actor *Customer* interacts with the *Waiting_Staff* component through the *Take_Order_Call* channel interaction, carrying the *Order* input parameter. Next, if the *Order* = *"Chinese"* condition is true then component *Waiting_Staff* shall interact with the *Chinese_Cuisine_Chef* component through the *Cook_Chinese_Food_Call* channel interaction; else if the *Order* = *"French"* condition is true then component *Waiting_Staff* shall interact with the *French_Cuisine_Chef* component through the *Cook_French_Food_Call* channel interaction; else component *Waiting_Staff* shall interact with the *Mexican_Cuisine_Chef* component through the *Cook_Mexican_Food_Call* channel interaction. Continuingly, if the *Order* = *"Chinese"* condition is true then component *Waiting_Staff* shall interact with the

Chinese_Cuisine_Chef component through the *Cook_Chinese_Food_Return* channel interaction, carrying the *Meal_1* output parameter; else if the *Order* = *"French"* condition is true then component *Waiting_Staff* shall interact with the *French_Cuisine_Chef* component through the *Cook_French_Food_Return* channel interaction, carrying the *Meal_2* output parameter; else component *Waiting_Staff* shall interact with the *Mexican_Cuisine_Chef* component through the *Cook_Mexican_Food_Return* channel interaction, carrying the *Meal_3* output parameter. Finally, actor *Customer* interacts with the *Waiting_Staff* component through the *Take_Order_Return* channel interaction, carrying the *Meal* output parameter.

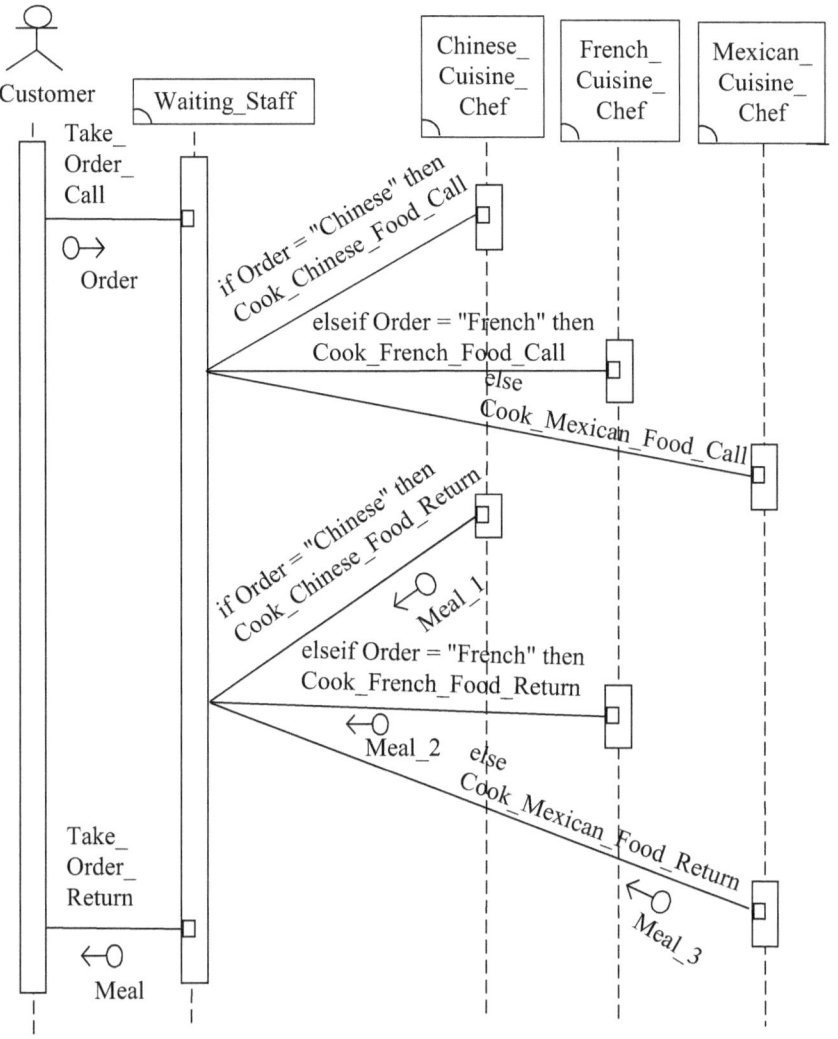

The channel-based IFD of the *Paying_the_Bill* behavior is shown below. First, actor *Customer* interacts with the *Cashier* component through the *Pay_Bills* channel interaction.

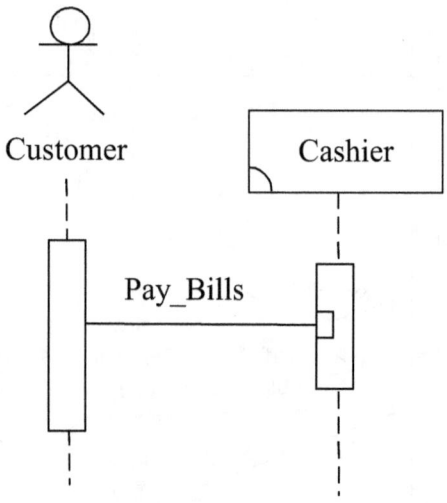

Operation-Based Systems Architecture of the Restaurant

The collection of all operation-based interaction flow diagrams defines the systems architecture. The overall behavior of the *Restaurant* includes two behaviors: *Ordering_a_Meal* and *Paying_the_Bill*. Each of them is described by an individual operation-based IFD.

The operation-based IFD of the *Ordering_a_Meal* behavior is shown below. First, actor *Customer* interacts with the *Waiting_Staff* component through the *Take_Order* operation call interaction, carrying the *Order* input parameter. Next, if the *Order* = "*Chinese*" condition is true then component *Waiting_Staff* shall interact with the *Chinese_Cuisine_Chef* component through the *Cook_Chinese_Food* operation call interaction; else if the *Order* = "*French*" condition is true then component *Waiting_Staff* shall interact with the *French_Cuisine_Chef* component through the *Cook_French_Food* operation call interaction; else component *Waiting_Staff* shall interact with the *Mexican_Cuisine_Chef* component through the *Cook_Mexican_Food* operation call interaction. Continuingly, if the *Order* = "*Chinese*" condition is true then component *Waiting_Staff* shall interact with the *Chinese_Cuisine_Chef* component through the *Cook_Chinese_Food* operation return interaction, carrying the

Meal_1 output parameter; else if the *Order* = *"French"* condition is true then component *Waiting_Staff* shall interact with the *French_Cuisine_Chef* component through the *Cook_French_Food* operation return interaction, carrying the *Meal_2* output parameter; else component *Waiting_Staff* shall interact with the *Mexican_Cuisine_Chef* component through the *Cook_Mexican_Food* operation return interaction, carrying the *Meal_3* output parameter. Finally, actor *Customer* interacts with the *Waiting_Staff* component through the *Take_Order* operation return interaction, carrying the *Meal* output parameter.

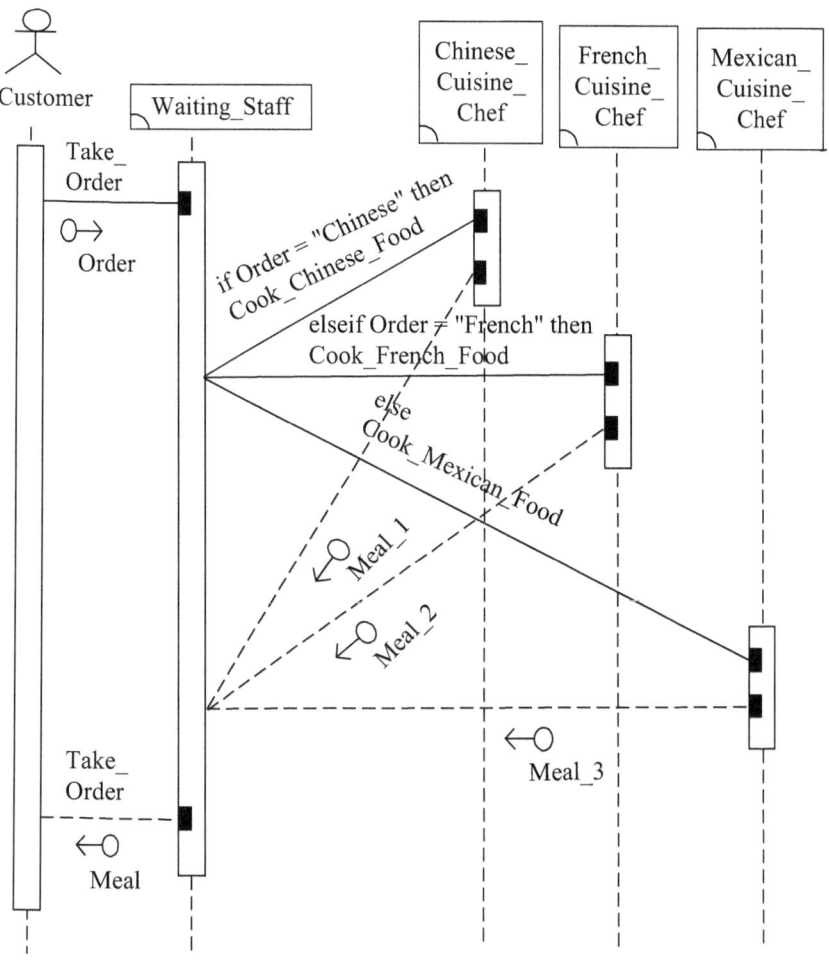

The operation-based IFD of the *Paying_the_Bill* behavior is shown below. First, actor *Customer* interacts with the *Cashier* component through the *Pay_Bills* operation call interaction.

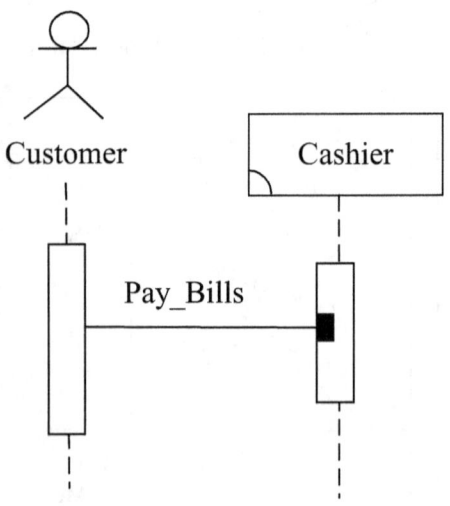

www.ingramcontent.com/pod-product-compliance
Lightning Source LLC
Chambersburg PA
CBHW070847180526

45168CB00002B/987